LEARJETS

Geza Szurovy

Motorbooks International
Publishers & Wholesalers ®

First published in 1996 by Motorbooks International
Publishers & Wholesalers, 729 Prospect Avenue, PO Box 1,
Osceola, WI 54020-0001 USA

On the Front Cover
Learjet 35A over San Francisco Bay. *Learjet/Paul Bowen*

On the Title Page
Learjet 35A. *Learjet/Paul Bowen*

On the Back Cover
Learjet 31.

On the Frontispiece
Learjet 60.

On the Acknowledgments Page
Author portrait—in the Learjet 60.

Library of Congress Cataloging-in-Publication Data Available

Szurovy, Geza
        Learjets / Geza Szurovy.
            p. cm. --(Motorbooks International enthusiast
        color series)
        Includes index.
        ISBN 0-7603-0049-6 (pbk. : alk. paper)
        1. Lear jet aircraft--History. I. Title. II. Series:
        Enthusiast color series.
        TL686. G37S98   1996
        629. 133' 349--dc20   95-26551

Printed in Hong Kong

CONTENTS

ACKNOWLEDGMENTS

This book would not have been possible without the unreserved cooperation of many people at Learjet. I would particularly like to thank Jeff Miller, Pete Reynolds, Don Grommesh, Dick Etherington, Frank Mastin, Bob Agostino, and Carolyn Schlegel for sharing their experiences, insights, anecdotes, and their love of the Learjet. Thanks also to the efforts and enthusiasm of my editor, Mike Haenggi, without whose support this book would have remained only an idea. Although they and others have been generous with their comments, any errors and omissions are mine.

A note on the Learjet name: it has been written in different ways over the years, starting as Lear Jet. For consistency's sake, it is written in its current form, Learjet.

A few observations about aircraft performance: it is always tricky to select the measures of aircraft performance. The great variations in performance based on many variables make generalizations difficult, yet some aggregate idea of performance parameters can serve as a useful benchmark. Unless otherwise stated, I have used the manufacturer's published performance figures. Range is generally maximum range with four passengers and a 45 minute fuel reserve. Aircraft are considered to be "transcontinental" if they can routinely make the crossing in both directions under a wide variety of weather conditions and loading configurations. Balanced field length is the distance within which an aircraft can accelerate to critical engine failure speed (V_1) and then stop or accelerate to a safe climb speed (V_2) and clear a height of 35 feet with one engine.

Geza Szurovy

One of the first color air to air photographs of Learjet 23-001, shortly after the first flight.

CHAPTER ONE

CHARGE!

First came the whine of turbines spooling up. Then the speck at the end of Runway 19 at Wichita Municipal Airport began to move, accelerating rapidly. After a short ground roll it lifted effortlessly into the dusk's fading salmon sky, a tiny silver bullet of an aircraft with swept wings, tip tanks, and a sleek T-tail, conveying a rakish elegance decades ahead of its time. The date was October 7, 1963, and the aircraft was the Learjet Model 23 (N801L) on its maiden flight. Under the guidance of test pilots Bob Hagan and Hank Beaird the flight went perfectly. The Learjet

The first Learjet is rolled out, surrounded by the people who built it.

was on its way to becoming the defining icon of executive aviation, just as its brilliant and mercurial creator, Bill Lear had promised.

William Powell Lear was 61 years old when he took the first serious steps to develop the Learjet. He had never designed an airplane before and wasn't an

aeronautical engineer. He had, however, worked in aviation his entire life as a self made electronics engineer and entrepreneur. Among his early inventions was a radio compass, and at age 47 he had won one of aviation's most prestigious awards, the Collier Trophy, for his invention of the first jet autopilot, the

First flight of the first Learjet, October 7, 1963.

F5. And by the 1950s his avionics firm, Lear, Inc. had made him a millionaire.

Lear had also been a pilot since 1922, and being an impatient dynamo of a man, was constantly dissatisfied with the ponderous aircraft available for executive transportation. His critical views led him into his first aircraft venture, the modification during the early 1950s of Lockheed Lodestars, then being widely used as executive transports. Teaming up with two talented aeronautical engineers, Gordon Israel and Ed

Swearingen, Lear managed to increase the Lodestar's cruise speed from 200 to 290 mph. But just then the first jet transports began to appear, and Lear's discontent with the typical executive aircraft of the day turned to despair. Cutting edge airliners were whisking grandmothers to family reunions from coast to coast in the stratosphere at near bullet speed while executives had to troll around in the clouds below 10,000 feet at a snail's pace, he complained. Executive transport had to be brought up to the standards of modern air travel, and he would be the one to do it. He flirted briefly with pusher turboprop power, an idea inspired by Dr. August Raspet's experimental Marvel developed for the US Army and one to which he would return late in life, but soon he decided that nothing less would do than a six to seven seat executive jet that matched or exceeded jetliner performance.

Lear wasn't alone in his efforts to develop a business jet. There would be other executive jets on the market by the time his creation would fly, such as the Saberliner, the JetStar, and in Europe the MS760, the DH 125, and the Mystere 20. But several of these aircraft were large, civilian versions of military VIP transports (royal barges, said Bill Lear), up to twice as expensive as the Learjet would be, and none delivered the speed promised by Lear. The Learjet would race only one other aircraft with similar specifications to certification, the Jet Commander.

By the mid-1950s, Bill Lear was living in Switzerland, expanding Lear, Inc.'s avionics business into Europe. As his ideas for an executive jet gelled, he was introduced through his son, Bill Jr., to the Swiss P.16 fighter-bomber and its designer, Dr. Hans L. Studer. Five P.16s had been built in an attempt to develop a successor to the Swiss Air Force's Vampires and Venoms, but two had crashed (both into lakes, leading local wags to call the P.16 the Swiss submarine) and the production order was canceled. Bill Jr. had flown the P.16 and was smitten by it. Lear was also sufficiently impressed by the airframe to hire Studer and his design team to develop the executive jet. Lear was of the opinion that the fundamentally excellent P.16's crashes and grounding were caused by poor systems design. The Swiss American Aircraft Corporation (SAAC) was set up in Altenrheim, Switzerland and design work got underway in earnest.

It has been widely believed over the years that the Learjet was a P.16 with a passenger fuselage. As much as Bill Lear himself liked to spread the notion at times that the Learjet was essentially a Swiss fighter in civilian clothing, this is not entirely true. While there are great similarities between the wings of the two aircraft, the Learjet does not have common components with the P.16. Rather, it benefited from the basic wing design research done for the P.16, with which it shares the laminar flow airfoil also found on the P-51 Mustang. No high speed wind tunnel tests were done for the Learjet. Instead it relied on the high speed wind tunnel tests done for the P.16 with the tail off, hence the similar wings.

In funding the design process, Lear was following his established custom of developing the idea on his own and then convincing Lear, Inc.'s board of directors to finance the prototype and production. But with the Learjet he was running into a problem. The board wanted to have nothing to do with the airplane. As far as the board members were concerned Lear, Inc. was not in the aircraft manufacturing business. Lear was sure he could convince them to go along once the design took shape, as had been the case with previous schemes, but this time the board wouldn't budge. Lear then simply resigned as chairman, sold his shares in the company for $12 million, and went back to Switzerland with his money to develop the Learjet on his own.

In Altenrheim the design was slowly taking shape and the aircraft was tentatively called the SAAC 23 Execujet, but Lear was encountering another problem. In addition to Dr. Studer and his Swiss staff,

One of the first Model 23 instrument panels.

Bill Lear demonstrates the strength of the Learjet's windshield.

Lear had hired a number of American engineers over time, including Gordon Israel of the Learstar project. Friction was developing between the Swiss and Americans, caused in part by personality clashes between Studer and Israel and in part by differing work styles. The Swiss were used to a slow, methodical work pace, normal working hours and weekends at home. They were consummate draftsmen who turned out museum quality renditions of working drawings and spent hours waiting for the ink to dry. Lear had other ideas. His timeline for the project was based on 12 hour days six or seven days a week and he didn't give a hoot if there was an ink spot or two on the plans. The Swiss also found it more difficult than the Americans to comprehend Lear's disorganized, freewheeling work style that could bring major changes at any moment. A British consultant concluded that at the pace the project was going it would fail unless SAAC was reorganized. Lear's response was swift. Practically overnight and without consulting anyone he announced that he was moving the project to the United States. He chose Wichita as the new location and in August 1962 he changed the name of the company to Lear Jet Corporation.

By January 1963 the Lear Jet Corporation completed the move into its new premises at Wichita Municipal Airport. Hank Waring, who had signed on in the final stages of the Swiss episode, was the new chief engineer, everyone understood the work pace required, and the project was back on track. "Charge!" became the company motto and the name of its newsletter. But Wichita's traditional, conservative aircraft manufacturers were skeptical of the upstart venture. In their estimation it would take an experienced aeronautical engineer ten years and $100 million to do what Bill Lear was proposing. It was time to prove them wrong.

As actual construction of the aircraft was about to begin, Bill Lear reviewed his financial resources and concluded that he had insufficient funds to follow the traditional practice of first custom building the prototype, then doing the flight testing to work out any design errors, and only then setting up the expensive tooling required to put the airplane into production. For the money to last through certification there would be no prototype. The very first Learjet would be built with the production tooling. Major design shortcomings would be ruinously expensive to fix. "With this approach you are either very right or very wrong," said Lear. He was convinced that he was very right.

Construction progressed rapidly. By June 1963, the fuselage forward and aft sections, as well as the wings were essentially completed and ready for mating. The first flight was only a few weeks away, when a design issue surfaced. The tail design had originally been a T configuration, but had been changed to a cruciform design in Switzerland. Subsequently, Hank Waring and chief of structures, Don Grommesh, had concluded that as configured, the horizontal stabilizer was too close the jet wake and the tail could suffer sonic fatigue as a result. Waring and Grommesh recommended a return to the original T-tail configuration, moving the horizontal stabilizer further away from the thrust line. Lear was unenthusiastic. He was impatient to get the Learjet in the air and the redesign would be a major extra expense. But he listened, as he always did when someone had something sensible to say, and agreed to the change. When the revised tail section was unveiled after a month an a half of non-stop shifts, he took one look and exclaimed, "That's the best-looking piece of tail I ever saw."

On August 16 the wings and fuselage were mated and on September 15, only nine months after completion of the move to Wichita, Learjet 001 was rolled out. It looked stunning. The tail redesign turned out to be a major contribution to its timeless, futuristic appearance, and a landmark aesthetic improvement over the cruciform tail.

Underneath the Learjet's sleek lines was an aircraft designed according to the conventional technological standards of the day, but with a practically evangelical commitment to maximizing performance. Efforts to minimize drag centered around several key features. The laminar flow wing, using the NACA 6 airfoil, also used on the early military jets, was a very efficient high speed

The Swiss P.16's design strongly influenced the development of the Learjet airframe.

airfoil. Although its stall characteristics were rather abrupt, its high speed performance was considered worth the trade off. Another NACA technique, the conical camber was employed to reduce stall speed, though it didn't improve stall characteristics. Area ruling, developed by Dr. Whitcomb of NACA during World War II to reduce transonic drag on jets was employed. Area ruling gave the fuselage its coke bottle shape and kept the two engines so close to centerline that the yaw effects of an engine failure were minimized. NACA data was also used to design the tightly cowled engine nacelles which were developed by the engine maker, General Electric.

The wing was exceptionally strong, built around eight wing spars. The design incorporated fail safe features, which allowed the in-flight failure of a major component such as a spar without incapacitating the airplane. Lear was also obsessed with keeping down

weight, and exerted considerable personal effort to lighten every component to the maximum extent possible. After he once said that he would sell his grandmother to save a pound of aircraft weight, shop floor jargon substituted "grandmothers" for "pounds."

Another major performance factor was the Learjet's set of two General Electric CJ610-4 engines. With 2,850 pounds of thrust each, they provided a phenomenal thrust to weight ratio of 1 to 2.19. The engines were the civilian variant of the military J85 engine used in the Northrop T-38 Talon and the F-5 fighter.

Lear was particularly proud of two innovative design features, the windshield and the door. The two-piece, 270 degree view, wrap-around, acrylic windshield, made by a specialty subcontractor, was an aesthetic touch unrivaled in grace by any other jet transport to date. Its development posed a challenge

Learjets meet two P.16s in Switzerland during a European tour by Bill Lear and friends.

given the forces it had to withstand compared to the more common sectional windshields that look like bird cages. Lear kept a windshield in his office and delighted in jumping up and down on it and dropping a cannon ball on it at the slightest excuse to demonstrate its strength.

The lightweight, outward-opening, two-piece, clamshell door was another departure from the standards of the day. Doors of pressurized aircraft opened inward at the time and pressed against the fuselage to minimize the risk of decompression. Their drawback was that they were heavy. Lear's extensively tested door resulted in the "sale of 150 grandmothers" per unit compared to the conventional option and has never experienced an in-flight hinge failure to date.

The maiden flight went like clockwork. With Bob Hagan performing the takeoff, the Learjet lifted off in about 1,880 feet, and climbed at 170 knots to 5,000 feet, where the gear was cycled. Then came a climb to 10,000 feet during which aileron response was checked and the rudders were pulsed. Low speed handling was checked at 110 knots clean and 100 knots in the landing configuration. Spoiler checks were next at 125 knots, followed by a 250 knot descent and low pass over the airport. Hank Beaird completed the first flight with a perfect landing. He laconically remarked that the flight had gone better than expected and he had expected a good flight. In his post-flight report Hagan wrote, "Lear Jet Corporation has itself one hell of an airplane."

The flight test program swung into high gear and resulted in a number of changes, chief among them a recontouring of the wing leading edge to improve low speed handling. The high speed airfoil was extracting a toll, and although the recontouring would put low speed handling within acceptable limits, it would not be considered a strong point of the airplane until resolved by the Softflite program years later.

Another change at this stage was a realignment of the tip tanks. They drooped, and the more Lear looked at them the less he liked their visual effect. Finally he had them realigned to be level with the wings for purely aesthetic reasons, even though the change slightly increased aerodynamic drag.

As summer approached, the flight test program was going well, Learjet 002 and 003 were also flying, and certification was almost in the bag. And a good thing too, because Lear had not only run out of capital but had also gone heavily into debt. The Lear Jet Corporation was winging it on customer deposits. Then came another potential setback.

On the morning of June 4, 1964, a black plume of smoke rose from a cornfield off the end of Wichita Municipal's Runway 19. Learjet 001 had crashed. There were some very tense moments and visions of bankruptcy at the Lear Jet plant until it was established that neither pilot was hurt and that the cause of the crash was pilot error. On a single-engine takeoff, the FAA test pilot and one of Lear Jet's own test pilots had forgotten to retract the

Bill Lear straps in for his first flight in 23-001.

spoilers and the airplane had settled into the corn field. The gear had sheared off, but otherwise the airplane was practically undamaged. Unfortunately, after the pilots scampered to safety, jet fuel spraying from a severed fuel line caught fire and 001 turned into a heap of ashes.

Everyone expected Lear to go on a rampage, but he was surprisingly cheerful. "Okay, we've just sold our first Learjet," he said. The insurance money from 001 would pay for completing the certification program with the remaining Learjets. Lear also wanted to squash any rumors about the Learjet's demise that might arise out of the accident. The morning following the crash he jumped into Learjet 002 and flew it to the Reading Airshow then in progress, the most important US industrial airshow at the time. At an impromptu press conference he explained the reasons for the crash and then he personally narrated a 25 minute exhibition flight by the Learjet he had just flown in.

Flight testing resumed, and on July 31, 1964, the FAA awarded the Learjet it's coveted Type Certificate. The formal presentation occurred a few days later, when FAA Administrator Najeeb Halaby came to Wichita to personally hand over the certificate to Bill Lear. The ten months it took to attain certification from first flight was a new record. More importantly, the Learjet had beaten the Jet Commander to certification by four months.

There was a party in Wichita the day the Type Certificate was issued, and while Bill Lear was skinny

The Learjet's eight spar wing design, which was retained on all Learjets until the Model 45 which has a three spar wing.

An early Learjet Model 23 under construction. In the background is 23-001.

dipping, his shorts disappeared. They were found the next day, with "FAA certification" written on them, flying from the company flag pole.

Bill Lear and his crew had done it. In four years and for $12 million they had built the airplane that industry experts had said would take ten years and $100 million to build. With its Mach 0.82 maximum speed, 41,000 foot operating ceiling, and 1,500 nm range, it provided jetliner performance for five to seven passengers. It outran the competition and cost anywhere from $100,000 to $500,000 less. And almost immediately after its introduction it would become a synonym in the public's mind for executive jet.

Typical cockpit, Learjet 24D.

CHAPTER TWO

LEARJET SET

Learjet 24, the first aircraft ever to be certified to air transport category FAR 25 undergoing icing tests behind an A-26.

The day after the Learjet received its Type Certificate, the Lear Jet Corporation had a product already on the assembly line and 22 firm orders in hand. It was also broke. The solution to raising the $6 million required to go into production was to take the company public. Lear filed the necessary applications with the Securities and Exchange Commission, and set about outfitting Learjet 003 for its first customer. On October 13, 1964, Learjet 003 was registered as N200Y, and was delivered to Herb Hamilton's Chemical Industrial Corporation of Cincinnati and became the first Learjet to enter service as an executive jet.

The public offering was completed in November 1964 when Lear sold 550,000 shares in the Lear Jet Corporation at $10 per share, representing 40 percent of the company. By then there were firm orders for 62 Learjets and the assembly line got into full swing.

Orders continued to mount rapidly as corporations and the wealthy flocked to join the jet set. The Learjet flew faster and higher than just about anything the competition had to offer, and was certainly far more beautiful. Priced at $595,000 fully equipped, it was also one of the least expensive busi-

A late Model 25 in 1982 inbound at the end of a long day. *Learjet/Paul Bowen*

ness jets. Companies with far flung interests in obscure places that would tie up a senior executive's travel schedule for days now found the whole kingdom within reach in a few hours.

Lear demonstrated the possibilities himself by leaving Wichita mid-morning for Miami in a Learjet to give a lunch address at an important conference and flying back to Wichita to comfortably make a 4:00 PM appointment to address another conference. By the airline schedules of the day he would have had to leave Wichita at 4:30 AM to make his noon appointment in Miami and could not have returned to Wichita before 11:00 PM. He would have also had to use six different airlines with two enroute changes each way. This was the Learjet's strongest selling point. But the glamour factor wasn't far behind. Frank Sinatra was among the first buyers, and a steady stream of celebrities taking demo flights did much to hype the airplane to the general public.

On May 21, 1965, the Learjet demonstrated its coast-to-coast capabilities by establishing world speed records in its class both westbound and eastbound on a same day round trip flight between Los Angeles and New York. Total elapsed flight time for the 5,005 mile round trip was 10 hours and 21 minutes with only one fuel stop each way.

Shortly after the record flight, five Learjets stole the flight display at Reading Airshow. They began their demonstration with a military style formation fly by. Then the four wingmen peeled off in sequence and the lead aircraft made a 400 mph low pass, pulled up in an 80 degree climb and rolled out of sight, breaking every FAA regulation specified for the display, to Bill Lear's great delight.

As orders mounted, the company was profitable. Its shares, initially offered at $10 per share, were trading above $80 on the New York Stock Exchange. At the National Business Aircraft Association meeting in October, Lear announced the Learjet's first derivative, the Model 24, and a new design, the Model 40 Learliner, which was to be a large executive jet and regional airliner seating up to 28 passengers. The Lear Jet Corporation was on a roll but its success was about to be challenged. A week after the NBAA meeting the first fatal

N427LJ refuels in the Azores on its 1966 around the world record flight.

THE WORLD ACCORDING TO BILL LEAR

"Don't take a nibble. Take a big bite!"

"Strive for design simplicity; you never have to fix anything you leave out."

"If you put up half the money, you can make half the decisions!"

"I'd sell my grandmother to save just one pound of weight."

"Don't tell me it can't be done!"

"Let's do something different."

"The trick is to discern a market before there is proof that one exists."

"Could five hundred men have painted the Sistine Chapel?"

"Too soon is just right!"

Dyed water is run over the elevator to see if it could seep inside (and freeze at altitude), causing stability problems. This was an experiment in search of possible causes of two early and unexplained crashes of Learjets.

Learjet crash occurred, followed by several others over the next few months.

The aircraft involved in the first fatality had just been rewired because it was experiencing electrical problems. At the controls was the Lear Jet Corporation's own test pilot, Glen David. He had just departed on an evening flight from Detroit to Wichita and was passing through 25,000 feet in stormy weather when the aircraft disappeared from the radar scope. No cause was ever established conclusively, but electrical failure was suspected, leading to spatial disorientation in instrument conditions; at the time Learjets didn't have a third, independently-powered standby gyro. Bill Lear also suspected a faulty oxygen valve that could have caused an explosion.

The first Learjet delivered to a customer. 23-003 about to be handed over to Chemical and Industrial Corporation of Cincinnati (October 1964).

Learjet 23-006 then...

The next crashes were attributable to pilot error. The hasty leap from slow, propeller driven piston aircraft was proving to be overwhelming for some pilots. One Learjet took off VFR at dusk in bad weather from Palm Springs with a pilot inexperienced in jets. While milling around at cloud base in the mountains, waiting for an IFR clearance, it flew into a ridge. One perfectly fine Learjet was put in Lake Michigan by its pilots (without fatalities). The first exported Learjet crashed in Zurich due to pilot error.

Then came another crash that couldn't be explained. A Learjet belonging to the Rexall Drug company (the third Learjet delivered) flown by highly experienced pilots in rain at 33,000 feet near Fort Worth, Texas, disappeared from the controller's radar screen. Again, no conclusive reason could be established and once more the NTSB listed electrical failure as the probable cause.

A general problem with a horizontal stabilizer hinge pin was also uncovered at this time—which was fixed on all Learjets in service and was later proved not to be a saftey issue at all—but it couldn't be linked to the crash. Lear then heard of water seeping into an airframe component of a DC 7, freezing, and causing serious stability

...and now. 23-006 is now on loan to the Kansas Aviation Museum in its original colors.

Note the engineering detail of the main gear of the Learjet 25D.

problems. He wondered if anything similar could have happened to the Learjet. He conducted an experiment and found that water could indeed seep into the elevators in flight. Furthermore, the pressure inside the elevators was lower than outside, trapping the water. Lear was convinced he had found the problem (although it couldn't be conclusively proven to have caused any of the crashes). A simple fix was designed and retrofitted to all Learjets. Concurrently the FAA mandated that a standby, independently-powered gyro be installed in all Learjets (a requirement in all turbine aircraft today). Whatever the reason, there were no more mystery crashes, and steps were also taken to improve pilot training. Learjet was back on track. The Learjet 24 received its Type Certificate on March 17, 1966. It was about to show the entire world the remarkable reach of executive jets by embarking on a record flight around the globe.

The Learjet 24 was essentially a Model 23 refined in minor ways. It was the first aircraft ever to be certified under the then newly introduced FAR 25 to airliner standard's which was a big accomplishment given the scrutiny it received being first. It had a stronger bird-proof windshield, a fire detection and extinguishing system in each engine nacelle, and an increased cabin pressure differential. It also had a redesigned electrical system that included an emergency dry cell battery that could run essential instruments in case of total electrical failure. There was an independently-powered standby gyro and beefed up brakes. To increase useful load, the gross weight was increased to 13,000 pounds from the 12,500 pounds to which Bill Lear had restricted the Model 23 in hopes of obtaining single pilot certification (which was not granted given the airplane's complexity and high performance). After the 104th Model 23, the assembly line switched to the Model 24. The company also offered a modification to upgrade Model 23s to Model 24s.

In mid-1966, Bill Lear learned from Arthur Godfrey that the actor was being given a Jet Commander to take on a round-the-world record flight. If Godfrey made it, the Jet Commander would become the first business jet to circle the globe. But not if Bill Lear could help it. He unleashed his formidable talents at making things happen fast. Within a week, and well before the Jet Commander was ready, a completely stock Lear Jet 24 blasted off from Wichita eastbound, to fly around the world. On board were pilots Hank Beaird, John Lear (Bill's second son), Rick King, and as official observer, Wichita journalist John Zimmerman.

The route they chose from Wichita would push the Learjet to the extremes of its capabilities on some segments: Bradley Field, Connecticut; St. John, Canada; the Azores; Barcelona, Spain; Istanbul, Turkey; Teheran, Iran; Karachi, Pakistan; Colombo, Sri Lanka; Singapore; Manila, Philippines; Osaka and Chitose, Japan; Shemya and Anchorage, Alaska; Seattle, Los Angeles, and back to Wichita. One anxious moment during the flight was the appearance of a MiG-15 off the Siberian Coast. The MiG shadowed the Learjet for some time before returning to Soviet airspace. The total trip took 65 hours 39 minutes and flight time was 50 hours and 19 minutes, setting 18 world records. It completely panned the Jet Commander and was a dramatic statement for the capabilities of business jets.

The Learjet was now solidly established as the leading business jet and work was progressing on a stretched version, the Model 25. Lear, getting bored with the routine operation of the company, began to make time to pour his creative energies into other projects. He had invented the eight-track stereo and now established a division to produce it. He bought Brantly helicopters, then a major civilian producer of two and four/five seat helicopters, and began tinkering with another pet project, a turbine helicopter. Design work progressed on the Learliner, and an order was placed for the Rolls Royce Spey engines that were to power it. The company was renamed Lear Jet Industries to more accurately represent its diversified interests. But the roller coaster world of aviation was about to take its next dive, taking Lear Jet Industries with it.

In the second half of 1966, an economic recession hit with a vengeance. In hard times, corporate fleets are among the first to suffer as executives temporarily revert to slogging it out in economy class on the airlines. Aircraft sales came to a crashing halt. Nothing moved. Not at Learjet, not at Jet

Soviet MiG 15 shadowing N427LJ on its around-the-world flight off the Siberian coast. This is no airshow! It is 1966!

Commander, nor anywhere else. Unsold Learjets were beginning to fill up the company ramp. The year-end results told the story. Aircraft sales registered a net loss of $12 million on sales that were down 50 percent from the previous year. The stock price fell to $8.50 from a high of $82. The only thing the company had going for it was an excellent product that was sure to bounce back as the recession wound down. But Lear didn't have the capital to ride out the cycle. Although there was modest improvement in early 1967 and the stock inched up to around $20, the only option was to sell the company.

Salvation came from the Gates Rubber Company, whose chairman, Charles Gates, had a pas-

sion for aviation. He was alerted to the opportunity by Harry Combs, from whom he had just bought Combs Aviation, a major FBO (fixed base operator) that Combs had founded and built-up over many years, and which included a Learjet dealership. Gates recognized a good product when he saw one and also saw a good fit with Combs Aviation which he planned to use as the Learjets' national sales organization. The deal was done, and although Bill Lear stayed on in the largely ceremonial position of chairman of the board for another two years, his role in the Learjet he had created with such passion was effectively over.

Shortly after the Gates acquisition, the Learjet 25 received its Type Certificate. It was essentially a simple stretch of the Model 24, providing two extra seats. The cabin was stretched 36 inches and the tail cone was stretched 16 inches which made room for another 400 pounds of fuel. The gross weight was upped to 15,000 pounds. In order to convey a new style following Lear Jet's acquisition by Gates, the large windows were replaced by the small airline style windows, which gave the aircraft a more streamlined look and also saved some weight. The small windows were also made available on the Model 24s at this time. Shortly after it was introduced in February 1968, a stock Model 25 blasted up to 40,000 feet in a mere 6 minutes and 19 seconds, taking away the time to climb world record in its class set by a Lear Jet 23 a year and a half before.

Under the Gates regime, Lear Jet Industries was renamed the Gates Learjet Corporation. It initially limped along and then slowly began to recover. The core aircraft division was separated from the other ventures which were folded into Gates Rubber, taking with them the great financial burden they had imposed on Learjet.

Financial recovery began in earnest when Gates convinced Harry Combs to come out of retirement and join Gates Learjet as President. Combs was in many ways as colorful a character as Bill Lear.

Combs instituted a program of tight cost controls he called "austerity with class, humility with pride." By the end of 1972, Gates Learjet realized a record net profit of $8.6 million. It had truly turned around and could look forward to a decade of prosperity and stability built not only on the series 20 aircraft but on a new turbofan that was to become the best selling Learjet of them all.

In April, 1976, the Learjet 24 and 25 accomplished another important first for Learjet when the Models 24E, 24F, 25D, and 25F all received FAA certification to fly all the way up to 51,000 feet. The supersonic Concorde was the only other civilian aircraft at the time certified to such an altitude. The project was undertaken to give Learjets the ability to take advantage of lower fuel consumption at that altitude, fly above the jet stream when desirable, and benefit from direct routings made possible by the scarcity of traffic up there.

Certification to 51,000 feet required a major redesign of the pressurization system, as well as improvements to the CJ610-6-engines which were prone to flaming out above 45,000 feet and were designated the CJ610-8 following the modifications. It was also a ground breaking experience with the FAA which had no certification standards to 51,000 feet, and worked closely with Learjet on this project to develop them.

The flight testing was among the more interesting missions flown by project test pilot Pete Reynolds and his colleagues. Other than having to contend with pesky engines that kept flaming out until the fixes were fine tuned, they had to fly all the way up to 51,000 feet, open a valve to depressurize the aircraft, and see what would happen as they conducted an emergency descent. Nor could they start down immediately. First they had to sit still for 17 seconds, the time the FAA decided it would take a surprised crew to react. "They were very long seconds," remembers Reynolds. During the most extreme tests the cabin rate of climb reached 60,000 feet per minute as the pressurized air inside the cabin rushed out, while the airplane was coming down at 24,000 feet per minute in its emergency descent!

Although the certification to 51,000 feet was a major technical accomplishment, operationally the Model 25 was too heavy to routinely fly at Flight Level 510. The 25D's could use FL470-490 under certain conditions (no Fs were built for customers). The lighter Model 24 could reach FL510, but few were built following the new altitude certification. The 51,000 foot certification program's true significance lay in what it would do for new Learjet Models in the years to come.

The 1,000th Learjet, a Model 35, in formation with a Model 23.

CHAPTER THREE

TURBOFAN

Learjet 25D and 36A off the California coast. *Learjet/Paul Bowen*

The idea of building a Learjet with fuel efficient turbofan engines was raised as early as 1968. The turbofan engine is a jet in which the jet turbine also drives a fan that is essentially a many bladed enclosed propeller. The turbofan engine converts more fuel energy into thrust than the pure jet, so it can produce the same amount of thrust on much less fuel. It is also much quieter than the pure jet. A difficulty with early turbofans was to get them to produce the power required at higher altitudes.

The initial concept in creating a turbofan-powered Learjet was to re-engine the Model 25 and call it the model 26. The model 26 was even announced at the 1969 Paris Airshow and the company went as far as to say that Model 25s would be retrofitable. Because of its more fuel-efficient engine, the Model 26 would be able to go further with the same load as the Model 25, or it could go the same distance with a greater payload.

Finding the right engine proved to be a major challenge. By this time Garrett—a Phoenix-based maker of small turbine engines—had developed two AiResearch turbofans, but one was too large and the other wasn't sufficiently powerful. Garrett's initial plan

had been to modify its Boeing 747 Auxiliary Power Unit, but by the time the engines were completed they bore little resemblance to the original APU.

Learjet asked Garrett to tailor an engine specifically to its needs. The specifications required an engine that could produce the same thrust at 41,000 feet as the GE CJ610. Learjet's spent a considerable amount of time convincing Garrett to build the engine, and the result was the TFE731-2, delivering 3,500 pounds of thrust. But there was a problem. When the engine was subjected to the chicken test (a frozen chicken blasted at it at several hundred miles per hour) to determine its ability to withstand bird strikes, the chicken won. Garrett's fix added so much weight to the engine that if it had been simply hung on the Model 26, it would have caused weight and balance problems under certain conditions. The fix was to stretch the fuselage by a foot, extend the wings by 48 inches, add more fuel, and name the aircraft the Model 35. According to Don Grommesh, the chicken's victory was one of the best things that ever happened to Learjet.

Learjet had to make a substantial financial commitment to Garrett for the engines. Business jet sales, however, were still lackluster. Cash flow was tight, and the obligation to Garrett ended up exceeding Learjet's net worth. The possibility of selling Learjet to Garrett was briefly raised. Garrett, however had no interest in becoming an aircraft manufacturer. Instead it took the engines and the testbed aircraft as security, and gave Learjet a cash flow boost by buying and paying in advance for the first production Model 35.

The prototype turbofan Learjet was a Model 25 on the production line that was first designated a Model 26 and was developed into the Model 35. A stock model 25 was also used for initial engine tests with the Garrett on one side and the original CJ610 on the other. The prototype first flew in August 1973, and easily met expectations. It was fully 35 percent more fuel efficient than the Model 25 and had a more comfortable cabin. Its maximum gross weight was

17,000 pounds, yet its light airframe (developed with Lear's traditional focus on weight control) gave it a generous useful load. With a typical range of 2,200 nm, it could fly farther than any previous Learjet. Its turbofans were noticeably quieter than the earlier generation jets, which was an important advance as strict noise regulations were coming into effect.

The Model 35 would have sold well in its own right, but it also benefited greatly from a political economic event outside Learjet's control, the global oil crisis of the early 1970s. Fuel prices had skyrocketed. An aviation fuel allocation program was put in effect in the US just as the Model 35 was coming on line, and the airlines were dramatically cutting back their route structure. A Learjet that was 35 percent more fuel efficient than its predecessors suddenly looked like a very wise investment. Learjet had 60 firm orders for the Model 35 even before it flew, and the order book climbed to 90 by the time the airplane was certified. Several corporations replaced a gas guzzling Gulfstream with two Learjet 35s to get the most out of the fuel allocation program.

Learjet also introduced a long range version of the Model 35. An extra fuselage tank was installed at the expense of two seats, gross weight was upped to 18,000 pounds and the airplane was designated the Model 36. It's typical range was pushed beyond 2,500 nm. The Model 36 could frequently accomplish nonstop coast-to-coast flights under most conditions and even transatlantic flights depending on wind conditions, especially eastbound. Taking advantage of favorable winds, a Model 36 with two pilots and two passengers completed a 3,411 nm non-stop flight from Hawaii to Wichita in early 1975, at the end of a two week global demonstration tour.

As America's bicentennial approached, Gates Learjet sought to demonstrate its turbofan equipped line's performance by organizing another round the world record flight in the spirit of the one flown by Hank Beaird and his crew in the Model 24 in 1966. Arnold Palmer, golfer, avid Learjet owner and pilot, and Gates Learjet spokesman, was selected to be the celebrity commander of the flight. He was accompanied by pilots James Bir and Lewis "Bill" Purkey of Gates Learjet and aviation author Robert Serling, who went along as official observer.

The Learjet to be used, a stock Model 36, was lent for the flight by none other than Herb Hamilton, the purchaser of the first Learjet ever sold. Hamilton also still owned the registration number which had been on that first Learjet, N200Y. Two hundred Yankee was the perfect N number for a flight to commemorate the bicentennial, so Hamilton had it temporarily reregistered to the Model 36. The airplane was named "Freedom's Way - USA," which was the winning entry in a national bicentennial slogan competition, and received a patriotic paint job for the adventure.

The route flown was Denver (site of the 1966 Aviation/Space Writers Association convention sponsoring the flight), Boston, Paris via a fuel stop at Galmorgan in Wales because of headwinds, Teheran, Colombo-Sri Lanka, Jakarta, Manila, Wake, Honolulu, and return to Denver.

It is interesting to compare the world record flights of the Model 24 and the Model 36, which came a decade apart. The average airspeeds were almost identical at 406 knots and 409 knots. However, Beaird and his crew needed 16 legs and had to fly 20,463 nm to complete the trip, while Freedom's Way made it in only 10 legs and 19,974 nm, shaving 1 hour and 31 minutes off the Model 24's flight time. The total elapsed time advantage in favor of Palmer's flight compared to Beaird and company was 8 hours and 13 minutes, accounted for by the fewer stops the Model 35 needed and the 3 hours the Model 24 had been forced to wait in Colombo for the headwinds to diminish. The trip's longest leg for Freedom's Way was 2,914 nm from Honolulu to Denver, twice as long as the longest leg flown by the Model 24 between Colombo and Singapore. Perhaps the most telling comparison is the fuel consumption on the two flights. Although the Model 36's flight time was only 1 hour and 31 minutes less, it consumed only 8,041 gallons of fuel compared to over 11,000 gallons burned by the Model 24.

Following the introduction of the 30 series aircraft, Gates Learjet made significant progress on a long term project to improve the performance and handling of its entire product line, including the 20

Learjet 35A. Note the sleek trademark windshield.

series aircraft. The image of Learjets as "hot ships" had persisted and the competition was making inroads with claims of lower stall speeds and more docile low speed behavior.

The company's first step in countering the attributes that had earned the Learjet its hot rod reputation was aimed at low speed performance.

Known as the Century III program, its objective was to reduce stall speed to provide lower takeoff and landing speeds and reduced balanced field length requirements. This was accomplished primarily by increasing the thickness of the wing's leading edge, thus producing more lift. However, while stall speed was reduced, the aerodynamic stall characteristics were unchanged and the stick shaker/pusher was retained. The stick pusher was modified to sense not only angle of attack, but also the rate at which the angle of attack is changing. As a result, the system was able to activate the stick pusher at lower angles of attack during high entry rate stalls but not during normal entry rate stalls. This allowed the pusher to be set closer to the aerodynamic stall than the original configuration, adapting the pusher to the reduced approach speeds made possible by the thickened leading edge contour. To further reduce runway length requirements the Century III program also included new high energy brakes.

The Century III 30 series Learjets, which also received avionics upgrades, became the Model 35A and 36A.

The Century III program was followed in 1979 by the Softflite program, which was aimed at improving handling characteristics in stall as well as at high speed. The initial flight testing of the Softflite program was performed on the Model 28 prototype (see following chapter), but the package was applied to all new production aircraft. It was also retrofitable to all previously manufactured Century III aircraft.

The Learjet's problem in stall was that the outboard section of the wing stalled first, blanketing the ailerons and causing an abrupt wing drop that could roll the aircraft beyond 90 degrees. Hence the mandatory requirement for the stick pusher. The Softflite program took two steps to tame the Learjet's stall. Stall strips were installed along the wing leading edge to trigger initial airflow separation in the stall on the wing's inboard section. Fences were installed parallel to the wing chord to restrict the inboard airflow separation from spreading outboard over the ailerons. As a result the wing stalled inboard first and progressed outboard gradually, eliminating wing drop in the stall.

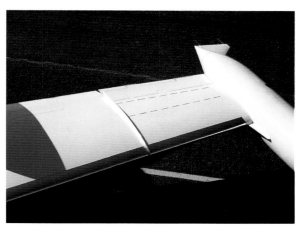

Century III wing with Softflite. Note the stall fence, boundary layer energizers. A stall strip is just visible on the leading edge to the left.

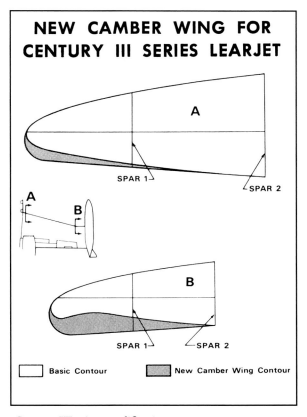

Century III wing modification.

Learjet 35A's interior was typical of the sophisticated executive style that became standard by the late 1970s.

In high speed flight, the objective was to delay the growth of the transonic shock wave over the wing. As Mach speed increases, the airflow at the thickest part of the airfoil approaches the speed of sound (Mach 1), causing a shock wave which begins to move aft. As the shock wave strengthens, the airflow behind it eventually cannot reattach to the wing. This transonic effect causes drag rise, buffet, aileron buzz, and longitudinal stability changes sometimes referred to as Mach tuck. Swept wings and thin airfoils delay the shock formation. An additional technique, initially used on Learjets, was lining the wing surface just ahead of the ailerons with vortex generators. The Softflite program replaced the vortex generators with lower drag boundary layer energizers (BLE's), which are essentially small flat strips attached to the wing surface. The onset of Mach buffet and aileron buzz are the same as with vortex generators, however, once developed, they are kept at a low level with the BLE's without any tendency to experience a wing drop. The BLEs, which were patented by Learjet, allowed a 50 knot increase in Vmo (maximum operating speed) to 350 knots indicated.

Softflite made the Century III Learjets so well behaved that in 1982, a Softflite I program was initiated to develop a modification which would bring similar characteristics to earlier Learjets. The solution to the undesirable stall characteristics of these "thin winged" aircraft proved to be a combination of the original Softflite program's stall fences and the addition of small metal triangles to the outer leading edge of the wings. These "shark teeth" disturbed the airflow at the stall in a fashion which caused both wings to stall simultaneously, eliminating the problem of a wing drop and resulting in a benign, straight ahead stall. All pre-Century III Learjets could be retrofitted with Softflite I.

The turbofan powered Learjet 35/36 went on to become the most successful Learjet to date. It was introduced at the right time and sold briskly as the demand for executive jets steadily grew throughout the 1970s. But the less expensive 20 series Learjet also remained in production and had made significant technical progress. And together with the Model 35/36 it formed the foundation for a whole new generation of Learjets.

ABOVE AND RIGHT
Thrust reversers became available on the Model 35.

Learjet 35 cockpit. Note pistol grip for braking parachute.

The Learjet 55B is easily distinguished from the later C model by the absence of delta fins. *Learjet/Paul Bowen*

CHAPTER FOUR

LONGHORNS

Rare portrait of 28-001, 25-064, and 24-218, the three aircraft that conducted the Longhorn flight test program. 24-218 is now reconverted into a Model 24 and flies with Calspan as a variable stability trainer for USAF and Navy test pilots.

At the 1977 National Business Aircraft Association convention in Houston, Texas, an intriguing new Learjet appeared. It looked like a Model 25 with extended wings but it also sported a set of graceful winglets. The Learjet was the Model 28 prototype and the winglets were the first ones ever on a jet and a production aircraft, civilian or military. The winglets were developed in great secrecy. Prior to the airplane's NBAA appearance, they were hidden by bogus tip tanks on the ground between test flights.

The design, dubbed the longhorn wing, was developed for the company's midsize Learjet project, the Model 55, which was announced the day after the Model 28's first flight at NBAA. Initially, the Model 28, which was the Model 25's fuselage and General Electric CJ610-8A engines with the new wing, was to be only the testbed for the wing. But it had such astonishing climb performance that Harry Combs, who loved to say that performance sold airplanes, decided it should be offered as a production aircraft in its own right.

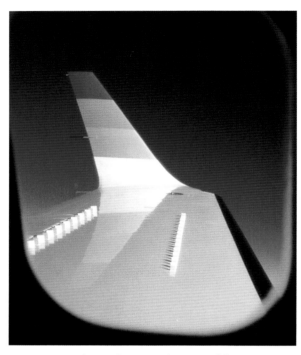

Learjet 28-001's winglet. Note the pre-Softflite wing with vortex generators instead of boundary layer energizers.

The impetus for the longhorn wing was the recognition that Gates Learjet needed a larger airplane with a stand-up cabin in its product line to compete in the mid-size market. The big challenge to Learjet was about to come from the recently announced Cessna Citation III (then still on the drawing boards), which was being heavily marketed as offering Learjet performance with a stand-up cabin. The days when Bill Lear could brush off complaints about a lack of headroom by saying, "You don't stand up in a Cadillac," and "If you want to take a walk, go to Central Park," were gone. For a clearly identifiable segment of the market the standup cabin was a critically important element of the buying decision, one that Learjet could no longer afford to ignore.

One option considered was modifying the Model 35 fuselage by increasing the headroom by 15 inches. This idea was discarded when it became apparent that the resultant fuselage shape would have

looked awkward, the vessel would have been difficult to pressurize, and the airplane wouldn't have had the range to carry a sufficient number of passengers the requisite distance. An entirely new fuselage and a high performance wing to carry it was needed to meet the new model's mission specifications.

The new wing was derived from the Model 35 wing by removing the original "straight" two foot extensions that were originally added to the Model 25 wing to make the Model 35 wing, and adding six foot extensions that continued the wing's sweep. The higher aspect ratio made the wing more efficient, improving fuel economy and extending range. However, Learjet engineers saw an opportunity to further increase the wing's efficiency by the addition of winglets, which were then being studied by NASA in wind tunnel experiments, but which had never been tried on jets or any other large aircraft. The theory was that winglets would reduce the drag caused by the wingtips and would also generate additional lift in a forward direction similar to sails, enhancing thrust and reducing the fuel required to fly the same distance, or extending range on the same amount of fuel.

Working with NASA's results, Learjet developed the winglets and installed the new wing on a Model 25 and a Model 24 for the flight test program in addition to the first production Model 28. Later, test flights were made with and without winglets. Following the resolution of some buffeting problems, the results showed that the winglets not only increased range by 6.5 percent but also improved directional stability, enabling the test pilots to hold the aircraft level in a full rudder sideslip.

With a gross weight of 15,000 pounds (equivalent to the gross weight of the Model 25) but with the more efficient wings, the Learjet 28 proved to be one of the highest performance civilian airplanes in climb. It's balanced field length was only 2,520 feet and it was the first business aircraft and the first civilian aircraft other than the supersonic Concorde that could climb directly to 51,000 feet fully loaded and routinely operate at that altitude.

In February 1979 the Learjet 28's climb performance was forcefully demonstrated when astronaut Neil Armstrong, the first man to walk on the moon, and Learjet experimental test pilot Pete Reynolds flew

Neil Armstrong at the controls of a Learjet 28, lifting off with Pete Reynolds on the way to five world records for time to climb and altitude (February 1979).

one of the former Model 28 test aircraft to 15,000 meters (49,215 feet), an official FAI altitude level for record purposes, in 12 minutes and 26 seconds, establishing a world record. They also set four other world records that day for altitude gain and sustained altitude in two weight classes (3,000-6,000 kgs and 6,000 to 9,000 kgs). To symbolize how far aviation had come since Orville Wright's first 12 second hop in 1903, at Harry Combs' urging the flights were made from Kitty Hawk, North Carolina. The records

The Learjet 55 and 23 commemorating sixteen years of Learjet production in 1979.

stood for nine years before another Learjet 28 broke them. The latter Model 28s records still stood in the lower weight class as of 1995, and in the higher weight class it took a Lockheed U-2 reconnaissance aircraft to break them.

The Model 28s drawback without the benefit of tip tanks was a rather short range of around 1,300 nm with a 1,200 pound payload. The increased efficiency of the wing didn't make up fully for the decrease in fuel capacity (the model 55 would have a much larger fuselage tank, higher gross weight, and the Garrett turbofans). To increase the Model 28s range, Learjet simultaneously introduced the Model 29. True to the well established Learjet formula, it had a larger fuselage fuel tank at the expense of two seats. Its reach increased to around 1,800 nm with a 1,200 pound payload, but

this still fell quite short of the 35A's 2,200 mile typical range. Nevertheless, there was some niche interest from buyers with short range mission needs who wanted to take advantage of the airplane's phenomenal climb and cruising altitude capabilities. In all, five Model 28s and four Model 29s were built. Telling of their niche capabilities, two of the Model 28s and two Model 29s are on the Mexican registry, while the remaining two Model 29s are flown by the Government of India around the Himalayas.

The Longhorn Model 55 first flew on April 19, 1979. By the time it was certified in March 1981 its order backlog was running higher than the initial order rate had run for the company's most popular Model 35/36. The mid-size Learjet had found a ready market. The first aircraft was delivered the month fol-

Family portrait: Models 23, 25, 36, and 55 (from top to bottom) represent the evolution of the Learjet from 1963.

lowing certification, beating its rival, the Citation III, by about a year.

The specifications of the Model 55 were indicative of how far the Learjet had evolved from N801L. The fuselage was a brand new design of spacious proportions with the highly anticipated stand-up head room as well as a full airline style lavatory. The longhorn wings, while derived from the Model 25, were in many respects a new design. Powering the Model 55 was an enhanced version of the Model 35's fuel efficient Garrett AiResearch TFE 731 turbofan, the 731-3 delivering 3,700 pounds of thrust per engine. The gross weight was 20,500 pounds (compare this to the Lear Jet 23's 12,499 pounds), and typical range with four passengers and two crew members was equivalent to the Learjet 35's 2,200 nm range.

The Learjet 55 production line was set up in Tucson, Arizona, where the company had progressively moved a variety of functions during the Gates years, including some component manufacturing activity, the final completion of green aircraft to customer specification, delivery, and certain maintenance services (later even corporate headquarters moved to Tucson for a time, before returning to Wichita).

As the Learjet 55's service experience grew, it became obvious that many corporate users weren't simply treating it as a stand-up Model 35. They were using it for longer range missions, including extensive overseas travel. The need for more range was addressed by introducing options to add various extra fuel tanks at the expense of baggage space, passenger seating, and in the longest range versions, the coveted lavatory.

Learjets 55-001 and 55-002 during the flight-test program.

There were three Model 55 options offering increased range. The 55 ER had one extra fuel tank in the external baggage compartment in the tail cone, which increased its typical range to about 2,500 nm. The 55 LR was a 55 ER with an additional double walled Branson tank in front of the fuselage fuel tank. It's typical range was around 2,700 nm. The longest range option was the 55 XLR. It was a 55 LR with a second additional fuselage tank. It's range was about 2,900 nm. Even though it had four passenger seats, there was barely enough payload left to carry the pilot and copilot when it was fully fueled. Only one XLR was built, for a German businessman and his son who flew it between South America and Europe.

A variety of minor modifications were made to the Model 55 in the three years following its certification, aimed primarily at increasing useful load and improving takeoff performance at high density altitudes where the airplane fell somewhat short of traditional Learjet performance (hot, humid days, high altitude airports). The weight and balance envelope was expanded and maximum gross weight was increased in increments up to 21,500 pounds. Boundary layer energizers replaced the original stall strips on the wing leading edge. Automatic ground spoilers, which deploy on their own upon touch down, and automatic power reserve, which automatically increases power to the good engine in case of engine failure on takeoff, were added. The airplane was also equipped with more powerful brakes, larger axles, heavier tires, and new gear doors. These modifications slightly reduced balanced field length requirements and allowed the airplane to take off with higher loads at high density altitudes.

The first big change in the Model 55 came in 1986. It brought cutting edge avionics technology into the cockpit. The airplane was equipped with a five tube Collins EFIS (electronic flight instrumentation) suite and a new Collins autopilot and was renamed the Model 55B. While achieving significant reductions in weight and maintenance requirements, the early EFIS displays looked similar to the mechanical instruments they replaced, causing some pilots to wonder about the advantages of the new technology. But the moving map and navigation information display capabilities were impressive. There was also a noticeable reduction in the pilots' instrument scan because of the ability to display airspeed, reference speeds, decision height, autopilot status, and other information directly on the attitude display to a greater extent than it was possible on the mechanical instrument they replaced. These benefits soon turned even the most conservative crews into EFIS fans. And it was only the beginning. In the coming years, electronically displayed, software driven instrumentation would develop by leaps and bounds, filling the cockpits of future Learjets with banks of computer screens and greatly easing the pilots' workload.

By the time the Model 55B was introduced, work was also underway on a significant airframe addition, designed to aerodynamically circumvent any deep stall tendencies. These were the delta fins that were to be attached to the tail section of the Model 55C. But they wouldn't appear until 1988 when they would be unveiled simultaneously on the Learjet 55C and the new Learjet 31 as we shall see. The longhorn program proved to be a great success, producing a fine aircraft in the Model 55. It was a steady seller and firmly established Learjet in the mid-sized business jet mar-

The Model 55B's five tube Collins EFIS cockpit.

ket. But in its normal configuration the Model 55 still lacked transcontinental capability. As the various attempts to extend its reach indicate, Learjet acutely felt the need for additional range. The next logical step would have been to start the development of a more capable mid-sized Learjet drawing on the experience gained from the longhorn program. However, by the mid-1980s the financial fortunes of Learjet were poised for another decline. Twelve years were to pass from the introduction of the Model 55 in 1981 before true transcontinental capability would be achieved in the Model 60. But before we embark on the company's next roller coaster ride, let's pause to take a look at some of the more intriguing but lesser known applications of Learjets in a wide variety of Special Missions roles.

Learjet 55-002 undergoing icing test in formation with modified T-33s.

Finnish Defense Force Learjet (undergoing tests in the US, hence the US registration).

CHAPTER FIVE

SPECIAL MISSIONS

These Peruvian reconnaissance aircraft were Learjet's first military sale.

In January 1965, a Boeing engineer suffered a cerebral hemorrhage at Cape Kennedy, Florida. He had to be rushed to a specialist in Cleveland within hours if he was to have a reasonable chance to recover. Boeing contacted Lear Jet to see if an aircraft could be made available for the mission. Lear Jet in turn called its first customer, Chemical and Industrial Corporation. In less than 30 minutes three seats were replaced by a litter stretcher in N200Y, the first Learjet ever sold, and it was on its way. The stricken engineer made it to Cleveland in time. This was the first special mission ever flown by a Learjet. It ushered in a multitude of roles for the airplane that were to extend far beyond the missions of the typical executive jet.

The Learjet's high speed and operating ceiling, its range, its brutally strong airframe, and its economical operating costs in relation to alternative aircraft of similar performance are the main factors that have made it very attractive for a diversity of special uses such as medevac, high-altitude photography, space research, and a variety of air defense and training functions.

An early application was high-altitude photography. The typical civilian photo aircraft of the day was a ponderous piston powered military surplus gas guzzler restricted to an altitude of about 29,000 feet at best. The Learjet, however, could comfortably cruise at up to 45,000 feet at a much higher speed. These attributes had the potential for dramatically increasing mission productivity per flight hour because of the much greater area coverage. A Learjet could pho-

The Marquardt 101 target reeling system in action.

tograph 100 square miles per minute, far more than could be covered by alternative means. This advantage made it a cost effective competitor in the high-altitude photo market which served the need of numerous disciplines including cartography, geology, soil science, forestry, agriculture, hydrology, engineering, oceanography, and meteorology.

The Lear Jet Corporation worked with a subcontractor to develop its own camera mounts and related equipment, and several other companies developed mounts of their own. One popular solution was the temporary mount, a pod which replaced the lower half of the door and housed cameras fully

LEFT
The Learjet is eminently suited for military maneuvering.

accessible from the cabin. Following a photo mission it could be replaced by the original door in minutes, returning the airplane to normal configuration ready for service as an executive jet. Such flexibility, also applicable in other roles, became a strong selling point of the Learjet as a special missions aircraft.

Within the three years following the type's introduction, Learjets had completed mapping assignments in Indonesia, Zambia, Singapore, Sweden, Tanzania, Libya, Egypt, Sudan, and Yemen. They were also engaged in long term programs to map for the first time the remotest corners of Mexico and Argentina. The first military sale of Learjets, two Model 25Bs to the armed forces of Peru, was for the purpose of high-altitude mapping.

Chinese Learjet 36A equipped with sideways looking radar (SLAR) that can map 46,000 square miles per mission, up to 60 miles off the flight path. *Learjet/Paul Bowen*

Finnish Defense Force Learjet, special equipment operator's console.

A particularly exotic role for a Learjet is filled by the Model 24 that was once used in the 51,000 foot certification program and the development of the longhorn wing. Sold to the Calspan Corporation of Buffalo, New York, which is known for engaging in some of the most esoteric aspects of aeronautical research, it is loaded with fly by wire control components and serves as a variable stability trainer for US Air Force and Navy test pilots and engineers.

When the Model 35/36 entered service, it too became a popular high-altitude photo platform. In 1985, Learjet succeeded in penetrating a market not known for its acquisition of executive jets when it sold three Model 35A and three Model 36A high-altitude mapping aircraft to the People's Republic of China. Among these aircraft was a Model 36A equipped with sideways looking radar (SLAR) which could map 43,000 square miles per mission up to sixty miles distant from the flight path.

Learjets have also become the platform for a highly specialized oblique camera developed specifically for the air-to-air filming of high performance aircraft. Much of the air-to-air footage seen in airline commercials, and many air-to-air action scenes in feature films were filmed by this camera mounted in a Learjet operated by Clay Lacy. The camera's designer won an Oscar for his efforts.

The C-21A is used by the USAF primarily in VIP transport and transition-training roles. *Learjet/Paul Bowen*

Learjets have also played an important role in space research. Prior to the advent of satellites and the Space Shuttle, astronomers had difficulty studying the infrared band of the energy spectrum because atmospheric attenuation obstructed infrared energy. Rice University astronomers and NASA got around this problem by mounting an infrared telescope in the Learjet 23 (049) of NASA's Ames Research Center and flying it at 50,000 feet, where they were beyond the effect of atmospheric attenuation. At this altitude the unmodified 23's cabin held a pressure altitude of 10,000 feet, but oxygen masks were worn as a precaution.

The NASA Ames Research Center has been an operator of several Learjets since its acquisition of 23-049. Among their many missions was the training of Space Shuttle mission specialists prior to the first Shuttle Launch. The Learjet served as a platform to train mission specialists in the operation of infrared telescopes and other observation equipment, saving NASA a tidy sum by significantly reducing the need for on-the-job training on Shuttle missions.

As jets were increasingly used in the EMS role, a variety of medical equipment configurations were becoming available. At one end of the spectrum was the option of temporarily installing stretchers in any production Learjet. At the other end was the flying intensive care unit such as the specially outfitted Model 24D operated by the Swiss Air Ambulance.

In addition to its high speed, another advantage the Learjet had as an EMS aircraft was its exceptionally strong airframe, which could keep sea level cabin altitude all the way up to 30,000 feet. This was an important attribute, because for many post-operative patients who had undergone major surgery, it is critical to remain at a pressure altitude more or less equivalent to the pressure altitude at which they were operated, until their scars healed. An unhealed incision could burst with a rapid change in pressure. Prior to the availability of pressurized aircraft, patients in danger of experiencing such problems had to be flown within 500-800 feet of the ground in unpres-

Interior of the C-21A in MEDEVAC configuration.

surized air ambulances. The Learjet's ability to maintain very low pressure differentials between the ground and cabin up to very high altitudes gives it great flexibility in avoiding weather on pressure sensitive EMS missions.

The Learjet's speed and its high cruising altitudes that maximize the chances for direct routing have made it one of the most popular organ-delivery aircraft for transplant operations. Time is critical in transplants. From the moment a heart is removed doctors have four hours to make the transplant. A liver must be removed in four and a half hours and transplanted within six hours. On a typical organ transfer mission, the Learjet picks up the organ retrieval team from the hospital where the transplant is to take place and flies them to the donor. The team works at full speed to retrieve the organ and is rushed back with it to the waiting patient.

Another area in which Learjets have been especially successful throughout the world is in a variety of military roles. As defense budgets came under increasing pressure, the cost of using combat aircraft for target towing and threat simulation was becoming prohibitive, especially for the armed forces of smaller countries. The first to capitalize on the Learjet's potential in an air defense training role was Swedair, an innovative Swedish civilian contractor engaged in providing target towing services since the 1930s for the armed forces of Sweden.

Swedair acquired two Learjet 24s in the late 1960s and equipped them with external hard points capable of carrying 600 pounds on which it mounted a target towing mechanism and an ECM (electronic counter measures) pod. The Learjets replaced an aging Gloster Meteor jet and several piston-engined Douglas Skyraiders. They proved to be versatile target towing platforms, capable of simulating a wide range of intruders for the training of anti-aircraft gunners who lobbed 8 to 120 mm shells at the bright red targets dangling below the colorful yellow and blue Learjets. The targets could be extended anywhere from 150 feet to two miles below the aircraft, and the ECM pods would simulate the jamming capabilities of the intruders to throw the gunners off track. Mission altitudes ranged from 1,500 feet to 15,000 feet and an acoustic miss distance indicator provided the gunners with an instant score via a computer. Swedair also sold these services to the armed forces of Denmark and Austria. Another early user of a Learjet in the air defense training role was the Yugoslav Air Force, which had two production Model 25Bs modified similarly to the Swedair machines.

From these beginnings, Learjets went on to capture a major share of the air defense and ECM training market worldwide, including services for the US Navy and Air Force. The Learjet is capable of matching and even exceeding the normal cruise speeds of many front-line jet aircraft at a fraction of the operating cost (even supersonic military aircraft fly supersonic mission segments for only very brief periods of time) and has much longer endurance without refueling than most combat aircraft. Its airframe, which can sustain loads of up to 9 Gs, gives it the ability to easily fly military flight profiles. Learjets are now equipped with the Marquardt 101 target reeling system that allows a wide variety of air-to-air, air-to-ground, and height-maintaining targets to be towed and reeled out on as much as 30,000 feet of cable. Sophisticated electronics and radar packed into the cabin and chaff dispensers mounted on the hardpoints also allow Learjets to play state-of-the-art hide-and-seek electronic warfare games. At times the games can turn hazardous. A target towing Learjet was shot down in Taiwan at an airshow when the "defenders" miscalculated their aim.

Following the introduction of the fuel-efficient turbofans on the Model 35A/36A, the company decided to capitalize in the late 1970s on the Learjet's increasing special missions experience by developing a 35/36A explicitly for military special missions roles. In the early 1970s, national sovereignty was extended to 200 miles off the coastline and many nations found a need to patrol vastly increased territory. This was the initial role identified for the special missions Learjet, which was for a time marketed as the Learjet Sea Patrol. It carried a 360 degree sea surveillance radar, low-light level television video that could read and make available in real time the registration number of a ship; anti-submarine warfare sonobuoys and listening equipment; multispectral infrared scanners, LOROP (long range optical) cameras; HF, VHF, and UHF homers; SLAR (side looking airborne radar); and wing-mounted external stores including flares, smoke markers, chemical dispensers, and rescue pods.

The Finnish Defense Force was the first customer for a Learjet specifically outfitted by the company for a variety of special missions roles with a purchase of three aircraft. In their camouflaged paint schemes, the three Learjets perform sea patrol duties, tow targets, fly surveillance and photo reconnaissance missions, transport priority cargo, and also function as air ambulance and search and rescue aircraft. In addition to ECM and towing equipment, the Finnish Learjets also have a belly drop hatch, oblique reconnaissance windows in the forward cabin, and camera and air sample pods. Since the Finnish Defense Forces sale, military special missions Learjets in a variety of configurations have also entered service with the armed forces of countries on several continents, including the Japanese Self Defense Forces.

The biggest military sale of Learjets was the lease and subsequent sale of 80 Model 35As to the US Air Force beginning in 1983 under its Operational Support Aircraft program. Designated the C-21A and operated by the Military Airlift Command, the USAF Learjets perform a wide range of liaison, VIP transport, and priority cargo missions. They are also used to provide transition training for pilots assigned to MAC from USAF pilot training to prepare them for flying MAC's multiengine heavy transports.

In 1982, after years of military threat simulation and training missions, the Learjet went to war. A simmering dispute between Argentina and Britain over the Falklands erupted into open warfare when Argentina occupied the desolate islands about 400 miles off its coast in the South Atlantic. Britain despatched a task force to retake them and Argentine

Argentine reconnaissance Learjet 36A. This aircraft saw combat in the Falklands. On one of the missions it flew, its sister ship, T23, was shot down by a Sea Dart missile fired by the British destroyer, HMS *Exeter*.

military and civilian Learjets were to play an important role as the war progressed.

The Argentine Air Force's Air Photographic Esquadron I of Grupo 2 operated four Learjet 36As as photo reconnaissance aircraft based at Trelew. As the British landings got underway, the Learjets flew their first reconnaissance mission on May 25 over the Port San Carlos beachhead at 40,000 feet. They were detected by the British destroyer HMS *Coventry*, but before it could launch any of its Sea Dart missiles, the Learjets had streaked past, mission accomplished. Argentine fighter bombers sank HMS *Coventry* later that day. The Learjet's reconnaissance mission was successfully repeated two days later.

Following a lull in aerial activity caused in part by dismal weather, the Learjets were once again launched on a reconnaissance mission on June 7. Led by Lieutenant Colonel Rudolfo de la Colina, the squadron commander, they flew four abreast on parallel tracks at 40,000 feet at intervals of several miles to maximize coverage and minimize the time spent over target. On this mission, the incoming Learjets were again detected by the British air defense network, and the destroyer HMS *Exeter* launched two Sea Dart missiles at the intruders. One fell harmlessly into the sea, but the other one scored a direct hit on Colina's Learjet, sending the wreckage into the frigid waters below. Not having the benefit of ejection seats—which would have been the only hope for escape following a direct missile hit at the Learjet's high cruise speed—the crew perished. The remaining three Learjets returned safely to Trelew, and the conflict ended shortly thereafter.

Civilian Learjets were also used in the Falklands war by an Argentine outfit called the Fenix Escuadron. It was a unit of civilian aircraft conscripted, along with their pilots, by the Argentine armed forces for the duration of the conflict. The pilots were given temporary commissions and sent

Singapore Airlines' Learjet 31s are used for a variety of flight training missions. *Paul Bowen*

in harm's way. Many of the missions were routine logistics and transport flights on the Argentine mainland, but others were not so benign.

One type of mission was the decoy attack on which Learjets and (ironically) British Aerospace HS-125s were launched in military formation at the British fleet. The objective of their simulated attacks was to draw the defending British Harriers off the real attack being carried out by Skyhawks and Israel Aircraft Industries Daggers at the same time. Learjets were also used to do the navigation on combat missions for the Argentine Air Force's Daggers, which did not have the over-water navigation capability to fly the long distances to the Falklands and back.

Learjets haven't tangled with missiles in combat since the Falklands War, but they continue to find new special missions roles. The FAA operates several Learjet 60s as airways check aircraft, and several Model 31s are used by foreign governments in a similar function. A disappointment for Learjet was not having the Model 31A selected for the USAF Tanker Transport Training System program, which was awarded to Raytheon's Beechjet.

One of the more unusual recent roles for a Learjet is Singapore Airlines' use of its fleet of Learjet 31s. The airline has had the cockpits specially configured to resemble its Boeing 747-400 and Airbus A310 cockpits and uses the Learjets in a variety of training functions, including initial training for first officers and transition training from the first officer position into the captain's seat. The Learjets are also used to shoot extra landings by pilots assigned to the airline's long range routes, who perform only a handful of landings per month flying the line.

THE MANY SIDES OF BILL LEAR

With over 100 inventions to his credit, Bill Lear was one of the most versatile entrepreneurs of our time. Among his non-aviation accomplishments are the invention of a practical car radio, the eight track stereo, and a steam powered bus.

After he left Learjet, one project into which he poured his energies was designing the Learstar 600, another large business jet in the abandoned Lear Liner's league. The Learstar 600 design was acquired by Canadair with whom Lear went on to have a brief, cantankerous relationship. The Canadair Challenger was quite a different aircraft by the time it was built, but its origins are undoubtedly in the Learstar 600 design; ironic considering that Learjets and Challengers are now part of the same product line.

Bill Lear's last aviation project was the Lear Fan, under the auspices of Lear Avia, Lear's new company set up for the project. The Lear Fan was a composite turboprop twin pusher, on which the two engines were geared to drive one propeller. Lear first considered this configuration as an alternative to the Learjet. Bill Lear passed away on May 14, 1978, before the Lear Fan was finished. Three were built after his death, and Hank Beaird performed the first flight. The story is told that the Lear Fan design group was about to approve major changes after Lear's death. The vote was called, but before any could be cast a tremendous lightning bolt struck the plant. Without Lear's guidance the project got derailed and the Lear Fan never went into production.

Bill Lear (right) in earlier days with his radio compass.

CHAPTER SIX

TOUCH AND GO LBO

A rare photograph of Learjet 35-001 sporting longhorn wings. Note the absence of delta fins and the vortex generators of the pre-Softflite wings.

Gates Learjet had weathered the recession caused by the energy crisis in the early 1970s primarily because of the perfectly timed introduction of the fuel efficient Model 35/36. By the early 1980s, the company had been consistently profitable for over a decade, but that was about to change as the next recession set in. In 1981, Gates Learjet had made profits of $22 million on sales of $565 million and sold a record 138 aircraft. By 1983, profits were only $449 thousand on sales of $350 million which represented only 45 aircraft sold, mostly Model 55's. By the end of 1984, Gates Learjet had lost $11 million for the year; and in 1985, its losses were $23 million with only 33 aircraft sold.

The news would have been even worse, had not the Department of Defense (DoD) awarded Learjet a $175 million lease and logistical support contract in 1983 for 80 Model 35As (designated C-21A's) for the US Air Force's Operational Support Aircraft program. It is not unreasonable to say that Learjet would have most likely faced bankruptcy without the DoD order. The effects of the slowdown in executive jet orders were also eased somewhat by the aggressive marketing of Special Missions Learjets.

LEFT
Test pilots Pete Reynolds and Jim Dwyer following the Learjet 31's first flight. Reynolds became head of flight test for the entire Bombardier Aerospace Group.

Learjet's 55C and 31 in formation over the Learjet plant at their 1988 dual certification ceremony.

Learjet was not alone among the executive jet manufacturers in being affected by the recession, but there was an important difference. The other companies were owned by giant corporations with immense financial resources which could provide much needed cashflow in times of crisis. The Raytheon Company had bought Beechcraft in 1980, Cessna was owned by General Dynamics Corporation, and Gulfstream was owned by Chrysler Corporation. While these corporate parents would by no means write blank checks, there was assurance of some reasonable financial support to bridge temporary financial setbacks. The family owned Gates Corporation on the other hand, had few financial resources to spare for bailing out Learjet.

Harry Combs had retired in 1982, and the Gates Corporation was itself approaching a time of transition. Charlie Gates was nearing retirement, and there was little interest in aviation among the other members of the company's senior management. Given this state of affairs, coupled with Learjet's financial troubles, there were few options for Gates. In April 1986, Learjet was officially put up for sale, attracting interest mainly from several financial investment firms which were then thriving on Wall Street's latest leveraged buy-out (LBO) craze.

In an LBO, an investor borrows money to buy an underperforming company that has valuable assets. The investor profits from the LBO after repaying the borrowed money in one of the two ways: the investor either aggressively restructures the company to significantly improve cashflow (which leaves the investor with a valuable, moneymaking company), or the investor sells off the company's valuable assets for a quick one-time gain.

The LBO firms attracted to cash-strapped and debt-ridden Learjet saw more of an opportunity in aquiring it to sell off its valuable assets (the Learjet product line and the Combs-Gates FBO chain) than in restructuring and keeping it.

Learjet's problem in attracting investors interested in it for the long haul as a going concern was that the cash-starved company had not developed a new product since the Model 55 in 1981. While the Learjet name continued to command great respect, the competition was catching up. Cessna had survived the collapse of its light aircraft business by transforming itself into a major producer of business jets. It had thrived with its docile Citation, explicitly developed for customers transitioning to jets from turboprops who were wary of "hot" jets like the Learjet. Cessna virtually had this lucrative market to itself because of Learjet's refusal to "build down." Now Cessna was making inroads with more capable Citations and Beechcraft was also taking away Learjet customers with the Beechjet, an upgraded version of the Diamond business jet it had acquired from Mitsubishi. The uncertainty about Gates Learjet's future was also making potential customers wary and hesitant to buy Learjets.

While negotiations were progressing with various LBO firms, the company received one more big financial boost from the C-21A program late in 1986 when the US Air Force agreed to buy the 80 leased Learjets for $180 million. This transaction provided much needed cash flow, but it wouldn't solve the need for new products, nor would an LBO. That would come from the efforts of Learjet's dedicated employees, who had taken it upon themselves to work against great odds and uncertainty to revive product development. They would upgrade the Model 55 and build a new Learjet, the Model 31, which would become the best handling aircraft the company had ever built and would play a leading role in ensuring the Learjet's survival.

A development for which Gates Learjet Corporation had given the go ahead was the addition of delta fins to the tail section of the Model 55. Much of the work had already been done in fits and starts, proving the concept viable, but the company's cash flow problems had prevented completion and certification earlier.

The delta fins promised to be an effective aerodynamic foil to any tendency of the Learjet to get into a deep stall. The danger of a deep stall in T-tailed aircraft is that the turbulent airflow coming off the stalled wings could overshadow the horizontal stabilizer, rendering it ineffective and making recovery impossible. Several aircraft, though never a Learjet, had been lost

The first production Learjet 31. The type quickly gained a reputation for its excellent handling characteristics.

in deep stalls. Pete Reynolds had deliberately put Learjets into deep stalls and had had a few anxious moments before he recovered with power. On one occasion he even reached for the spin recovery parachute handle before convincing himself to give the airplane a few more seconds to recover on power alone.

Because of these characteristics, Learjets were required to be equipped with the conventional prevention device for deep stalls. This was a stick nudger-pusher calibrated to automatically push the stick forward and pitch the nose down well before reaching the angle of attack that would result in a deep stall. The delta fins offered an elegant alternative solution. They are essentially lift generating surfaces, which are in clear airflow below the T-tail when the tail surfaces are in the turbulence coming off the stalled wings. The fins are mounted at an angle to start generating lift when the wing stalls, thus lifting the tail and pitching the nose down. They worked like a charm mounted on Lear-

jet 55-002 and were put into production on the Model 55, which was designated the Model 55C.

Concurrent with the final development of the delta fins, the company was secretly working on the Model 31, contrary to an explicit ban of the project by Charlie Gates. Earlier, Learjet had experimentally mated the well behaved longhorn wing with Model 35-001's fuselage and Garrett turbofans and the results had been promising, yielding very pleasant handling characteristics. Enhanced with the delta fins and developed into a production aircraft, the configuration would at last provide an entry level Learjet to compete with the lower end Citations, and it would not be building down. Although the Model 31 would have less range than the Model 35, it would still be a Learjet with a hefty thrust to weight ratio and certification to 51,000 feet.

The Model 31 project was a true shoe-string operation. Experimental flight's Model 35-001 was turned

LEARJETS THAT NEVER WERE

Aircraft manufacturers consider many designs that never get built. Some are no great loss, but there are always a few that beg the question, "What if....?" In Learjet's case two proposed models that weren't built particularly stand out: the Model 40 Lear Liner, and project 7201.

The Lear Liner was to be Bill Lear's next big bite following the Learjet. It was going to be a large executive jet/regional airliner, to be powered by Rolls Royce Spey engines and capable of carrying up to 28 passengers. The project progressed all the way to the completion of the design by the summer of 1966. The production of the prototype was scheduled for the fall, but other projects used up Lear's money, and the Lear Liner was never built. Considering the subsequent success of aircraft such as the Gulfstream, the Falcon 50, and the Challenger/Regional Jet, Bill Lear may once again have been ahead of his time.

The 7201 project was proposed by Learjet's design engineers during the Gates years. It was to be an entry level business jet, designed to compete with the Citation, but Harry Combs quashed the deal because "Learjet would not build down." This decision left the entry level market to the Citation which went on to become one of the most popular business jets.

The Model 40 Lear Liner

Delta fins going on 55-002 in Learjet's experimental hangar.

into the prototype. Lately it had been used to develop Special Missions capabilities and sported the camouflage paint job which it had worn on demonstration flights. The stealthy fuselage was perhaps an asset, considering that Gates had forbidden the project.

The story is told that Charlie Gates visited Learjet's Wichita plant in connection with the proposed sale of the company. Great concern was caused by his scheduled stop in the experimental hangar where the Model 31 was progressing nicely in defiance of his wishes. It

was impractical to move the airplane, and covering it up might have aroused suspicion. In the end it was pushed as far into its corner as possible, the lights were turned off around it and all tools and accessories were removed. It looked like a dormant carcass. The Model 55 across the hangar floor, being fitted with the delta fins, was lit up like a Christmas tree and surrounded by an army of beavering workers. Gates is said to have remarked that the Model 28 was a fine airplane as he walked right past the Model 31 to admire the Model 55.

As the Model 55C and the Model 31 were about to make their first flights, Learjet found a buyer. In August 1987, Gates Corporation announced that it had agreed to sell Learjet to Integrated Resources, a New York LBO firm heavily invested in real estate and with no prior experience in the aviation industry. The price was $7.25 a share, well below the $10.00 per share Bill Lear got when he took the company public, especially when adjusted for inflation. But it was a reprieve of sorts, allowing Learjet to limp along, and the company made the most of the opportunity.

The Model 31 was a home run. It was so docile that neither stick pusher or Mach trim was required by the FAA, yet it had traditional Learjet performance. It was also gorgeous. With its range of 1,500 nm, ease of handling, and miserly fuel burn it was right on par with the entry level business jets, but it delivered so much more. No entry level competitor could match its power and speed, or its ability to fly routinely at 51,000 feet. The aviation press and potential customers were impressed. "Learjet Tamed!" ran the headline of one flight test article. The Model 55C was also well received. Learjet had a fighting chance.

The company also did what it should have done much earlier to cut costs by finally making the decision to reconsolidate aircraft production in Wichita. Orders started picking up, and in August 1988, Learjet made a dramatic announcement of its renewed commitment to the industry with a ceremony at its Wichita plant to celebrate the simultaneous FAA certification of the Model 55C and the Model 31. It was a great boost to employee morale to see the two aircraft in a formation low pass, though the company was by no means out of the woods yet.

Integrated Resources had made strong statements about having no intention of selling Learjet's Combs Gates subsidiary, "the crown jewel of the FBO industry" as a senior Integrated Resources official put it.

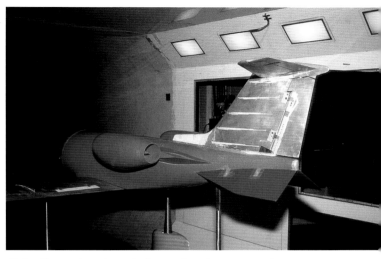

Delta fins undergoing wind tunnel testing.

There were even hints at possibly expanding the FBO chain. But Combs Gates was put up for sale soon after Integrated Resources completed its acquisition of Learjet, and in the first half of 1988, the FBO chain was promptly sold to the aviation services subsidiary of AMR Corporation, the parent company of American Airlines (Learjet would later end up buying back the four largest facilities of the FBO chain). Integrated Resources, however, was not to implement any further plans it may have had for Learjet. The bubble was about to burst in the real estate market, where the highly leveraged company had the bulk of its investments. By mid-1989, Integrated Resources was no longer able to service its massive debt, and soon thereafter it filed for bankruptcy. Learjet's future would now be up to the bankruptcy court, and the ability of Brian Barents, who had become Learjet's President in 1989 and would soon also be its Chairman, to get the best deal he could.

The Learjet 31 on its first flight. The camoflage paint scheme dates from the earlier use of the fuselage on a Special Missions demonstration aircraft.

A Learjet 31's wake turbulance parts the clouds on an early moring flight. *Learjet/Paul Bowen*

A Learjet 60 turning final with gear and flaps down.

CHAPTER SEVEN

A PERFECT MATCH

With Integrated Resources in bankruptcy, Learjet's future once again came into doubt. Would a competitor acquire the company and merge it into its own operations? Would it cease to exist as Learjet? The most logical suitor was not a competitor but an aircraft manufacturer with a product line which would be complemented by the addition of Learjet.

This Learjet 31A over a Louisiana bayou has an unusual paint scheme.

The first such company to show interest was Gulfstream, then owned by Chrysler Motors. Gulfstream builds only large, super-expensive executive jets which have earned it a high reputation for product quality not unlike the reputation of Learjets. Its acquisition of Learjet would have created an exciting partnership. Gulfstream went as far as to issue Learjet a letter of intent, but there was a surprise in store when the deal was presented to Chrysler's board of directors. Instead of approving the Learjet acquisition, the board voted to sell Gulfstream. The auto-motive industry was experiencing record losses, and sticking to core businesses was in vogue. Gulfstream was ultimately sold in a leveraged buy-out that was more successful than Learjet's tangle with Integrated Resources. But this turn of events did little to solve the problems of Learjet.

Then Learjet attracted the attention of Bombardier, a giant, low-key Canadian manufacturer based in Montreal with annual sales of over $3 billion in the early 1990s. Despite the suggestive name, Bombardier originally had nothing to do with avia-

The Learjet 31 is equally at home at small local airports as well as crowded airline hubs.

tion. When Bill Lear was creating the Learjet, Joseph Armand Bombardier, an equally entrepreneurial inventor and tinkerer, was making it big with his most lucrative invention, the Ski-Doo snowmobile. But by the time the company got interested in Learjet, it was deeply involved in the aviation industry. It owned Canadair, maker of the Challenger business jet and the CL-215 water bomber, and Short Broth-

ers, the Northern Ireland-based aviation manufacturer that had once built aircraft for the Wright brothers. It was also one of the world's premiere makers of railroad and subway cars. Snowmobiles amounted to only 10 percent of its business.

Bombardier's transformation from snowmobile maker to diversified transportation manufacturer was orchestrated by J. Armand Bombardier's son in law, Laurent Beaudoin, who took over the company when Bombardier passed away in 1964. The energy crisis that gave such a boost to the Learjet 35/36 was disastrous for snowmobiles, triggering big losses at Bombardier. Beaudoin recognized that to survive, the company had to diversify. Bombardier's first break was winning a contract to manufacture subway cars for Montreal under license from a French company. Then came a major order for subway cars from the city of New York. It allowed Bombardier to prosper as a shake out commenced in the industry, and gave it an opportunity to acquire weakened competitors. Beaudoin developed a well deserved reputation for being a shrewd deal maker. In the mid-1980s, his attention turned to the deals to be had in aviation.

Governments have always played a big role in the aviation industry as suppliers of lucrative defense contracts. Outside the United States, governments were also often direct owners of civilian and military aircraft manufacturers which were either acquired in government bailouts or owned from inception out of national interest. By the mid-1980s, two government-owned aircraft makers, Canada's Canadair, and Britain's Shorts, were experiencing serious financial problems. Having sunk billions of dollars into these firms, the governments were realizing that they were not the most efficient operators of these businesses, and wanted out. One option was to shut down the companies, but that would have meant a big loss of jobs. The other option was to privatize them. The underlying products were fundamentally sound, but the companies were sinking

The Learjet 60 demonstrating its excellent climb performance.

A Tranair solution plotting the airflow and local Mach numbers on a Learjet 60.

under massive financial obligations. If their financial slates were to be wiped clean they would become attractive to qualified private buyers. The jobs would survive and the product lines would have a good chance of achieving commercial success.

For a company like Bombardier, these privatizations were a great opportunity to acquire attractive assets at bargain basement prices and take on the challenges and risks of restoring them to profitable businesses. In 1986, the Canadian government wrote off over $2 billion on the Challenger program and sold Canadair to Bombardier for $121 million. In 1989, the British government injected $1.4 billion into Short Brothers, the largest single employer in Northern Ireland, to pay off debt and modernize the production facilities, and sold the company to Bombardier for $58 million. In short order, Bombardier would re-establish both companies as going concerns.

Bombardier's newly acquired Challenger was a direct competitor of the large executive jets made by Gulfstream and Dassault. The idea of acquiring a full range of executive jets to complement the Challenger was very attractive and Learjet was available. For Learjet, a sale to Bombardier would mean having at last the financial support and security that can come only from a well-run major corporation. It would be the perfect match. Beaudoin and Barents quickly came to terms, the bankruptcy court approved the transaction, and on June 29, 1990, the two companies announced the deal. Learjet would retain its separate identity with Brian Barents at the helm.

Under Bombardier's ownership, the reinvigoration of product development was made a priority. In October 1990, Learjet announced the new mid-sized, stand-up, walk-around cabin Learjet 60, to be powered by Pratt & Whitney's recently introduced PW305A engines delivering 4,600 pounds of thrust each. Simultaneously the company also unveiled the Model 31A, an improvement of the well received Model 31.

The main feature of the Model 31A improvement was a fully integrated Bendix/King digital flight control/avionics package which put electronic flight instrumentation systems (EFIS) into the cockpit. Dual digital air data computers (ADC) and dual attitude heading reference systems (AHRS) provided flight data to the five-tube EFIS 50 display and flight control information to the autopilot/flight director. The EFIS package was also fully integrated with the airplane's navigation systems.

Learjet put exceptional effort into working with Bendix/King engineers to fine tune the flight control software. As a result, together with its already excellent handling characteristics, the Model 31A became the smoothest Learjet ever, on or off autopilot.

The speed envelope was also expanded on the 31A, increasing Vmo from 300 to 325 KIAS, and Mmo from Mach 0.78 to Mach 0.81. This required the reinstallation of the Mach trim system, which was initially in the prototype Model 31, but was left out of the Mach 0.78 production models.

Other Model 31A improvements included digital nose-wheel steering, electric defogging for the windshield (instead of bleed air), and a more functional layout of the instrument and control panels and the circuit breakers.

The big project immediately following Bombardier's acquisition of Learjet was the mid-sized Model 60 destined to succeed the Model 55. Developing the

The Learjet 60 had the highest gross weight rating of all the Learjets at 23,100 pounds.

Model 60 as quickly as possible was an important goal. Thus, certain aspects of the basic design work were done concurrently with work on the prototype. In Bill Lear's time, such an approach was the standard, but by the time the Model 60 was going on the drawing board, the normal process was to fully complete all design work and only then build the prototype. Performing the two tasks concurrently in the interest of time entailed a certain element of risk to the schedule in case of a need for design changes. The schedule did come under pressure, but changes were accommodated in good spirit, and certification and the delivery of the first airplanes were accomplished by the set deadline.

The Model 60 was derived from the Model 55. The original intent was to stretch the fuselage and use the Garrett 731-5 engines. But when the Pratt & Whitney Canada PW305A engines became available, which were more powerful and had better fuel specifics, they were selected instead of the Garretts.

The Model 55-001 airplane was used as the prototype Model 60. It was cut in half and a two foot plug was added. This confirmed the need for more tail moment resulting in a 15 inch stretch of the tail cone. A relatively easy modification was repositioning the door to increase cabin room.

The stand-up cabin and lavatory aft are attractive features of the Learjet 60's luxurious interior.

The engine pylons essentially had to be redesigned to accommodate the new Pratt and Whitneys. There is a particularly complex three dimensional flow field around the pylons which presents a challenging task for aerodynamic engineers. For the first time, Learjet applied computational fluid dynamics to model the entire airplane and the airflow around it using NASA's Tranair (trans-sonic aerodynamics) software on a Cray computer at the NASA-Ames Research Center.

In addition to helping achieve the desired results with the new pylons, Tranair modeling also enabled Learjet to tweak the aerodynamics of the wing-fuselage fairings and the winglets. Pulling G's at altitude, the Model 55 wing had airflow separation at the winglets, creating a buzz. The plotting of local Mach numbers around the affected areas with the Tranair program revealed local flow problems, allowing Learjet to optimize the design of the Model 60 wing. The end result of using the Tranair program was a 4 percent reduction in cruise drag and a 40 nm increase in range.

The most significant technological advance in comparison to other Learjets was the Model 60's full authority digital engine controls (FADEC). FADEC is essentially power by wire. There are no physical control links between the cockpit and the engines. It is all done by software. FADEC had been in use on

The Learjet 31A's Bendix/King EFIS suite replaced the Model 31's tradition flight and navigation instrumentation.

airliners for several years before its introduction on the Learjet 60. Still, that first takeoff must have required quite a leap of faith in the cockpit.

Redundancy is a key feature of FADEC. There are two totally independent control units, each hooked up to different sensors. The system also has its own separate alternator to power the engine electronics. On startup the system goes through a self diagnosis, comparing the two units (channels). If there are any discrepancies the engines simply won't start. If all is well, the system automatically selects one channel and maintains the other one on standby. To make equal use of both channels, the standby channel is selected as the primary channel during the next engine cycle. The crew can also manually select a channel of preference.

With FADEC has come an even higher degree of cockpit automation. Starting each engine is literally a single flip of a switch. There are three power settings for the thrust lever: takeoff, maximum continuous power, and maximum cruise power. The pilot sets the appropriate detent and gets the required power, supplied by FADEC. The need for poring over power charts and fine tuning power settings is greatly reduced. Among other features, FADEC also sets automatic power reserve on takeoff in case of an engine failure and automatically reduces reverse thrust on landing rollout as the aircraft slows to reduce the chance of ingesting any debris on the runaway.

The Model 60's Collins Pro Line 4 avionics suite is also state of the art. There are four integrated displays. Each pilot has a primary flight display (PFD) and a multifunction display (MFD). The PFD contains in one single display all the flight and navigation information that pilots previously had to acquire through their traditional instrument scan. Among the wealth of information it displays is airspeed trend, telling the pilot where the airspeed will be in ten seconds based on current trends, and Vref (approach speed), which it calculates automatically for current conditions. As on the Learjet 31A, dual air data computers and attitude heading reference systems provide flight data to the displays and the autopilot/flight director.

When its flight testing got underway, the Model 60 flew as sweetly as the Model 31A, and with its powerful Pratt & Whitneys it was another rocketship in true Learjet tradition. Yet it also proved to be the most fuel efficient of the mid-sized business jets, with a typical maximum range of 2,750 nm, and true transcontinental reach with IFR reserves under a wide range of weather conditions and loading configurations. It was certified to a ceiling of 51,000 feet and could reach 41,000 feet in about ten minutes.

The Model 60 also delighted passengers. It is the most spacious Learjet with a cabin as luxurious and roomy as any the competition has to offer. There is also a full cabin width lavatory aft, generous baggage room, and a sophisticated electronic entertainment system including a flight information and moving map display and optional individual video screens for every seat. The company's efforts began to pay off handsomely when in 1994 the Learjet 60 became the bestseller in its category.

The Model 55C and the Model 31 kept Learjet in business during some of its most difficult days. Following the company's acquisition by Bombardier, the Model 31A and the Model 60 confirmed that Learjet was here to stay. With a solid, competitive product line complemented by the link to the Challenger, Learjet could now do what it hadn't done ever since the original Model 23. It could design and build a brand new aircraft from the ground up.

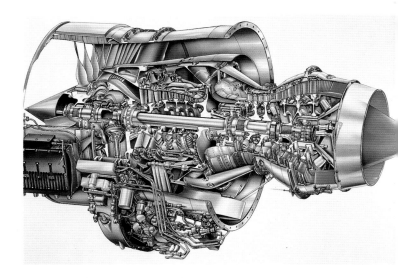

The 4,600 lb thrust Pratt & Whitney 305 engine powers the Learjet 60. It is controlled from the cockpit by full authority digital engine control (FADEC).

The way it is today. The Learjet 45 is a paperless airplane, designed entirely on screen.

CHAPTER EIGHT

THE MODEL 45

At the 1992 National Business Aircraft Association convention, Learjet announced the Model 45. For the first time since the Model 23, the proposed aircraft was to be a totally new design. It was also to be a cooperative venture of the entire Bombardier Aerospace Group, which by now also included De Havilland-Canada which was bought by Bombardier following its acquisition of Learjet. Learjet would have overall design and project responsibility for the Model 45. It would also build all systems and perform final assembly. Shorts would build the fuselage and De Havilland would perform the detailed design and construction of the wings.

The Model 45 was developed to follow in the footsteps of the Model 35. Its specifications called for an aircraft that had the performance of the Model 35, the handling of the Model 31A, and a passenger cabin more spacious than the competition. In normal configuration, the Model 45 seats eight passengers in a double club seating and has a full fuse-

The Model 45's aerodynamics are much easier to analyze on computers.

lage width lavatory aft. Its typical range is 2,200 nm, maximum cruise speed is Mach 0.81, and it is designed to fly all the way up to 51,000 feet. The powerplant is the 3,500 pound thrust Garrett TFE731-20, the most recent version of the veteran Garrett turbofan that

The Learjet 45 made quite an impression at its official rollout on September 14, 1995. *Paul Bowen*

has powered 30 series Learjets and the Model 55 going back all the way to 1973. This highly refined variant of the TFE731 series engine has better fuel efficiency and substantially higher cruise thrust at altitude.

Standard avionics include Honeywell's Primus 1,000 integrated avionics system. Flight and navigation information is displayed on four eight by seven inch EFIS screens and the panel also includes an engine instrument and crew alerting system (EICAS).

The Model 45's aerodynamic design was done using computational fluid dynamics. The analysis was done on NASA's Tranair software program, similar to the Model 60.

In another Learjet first, the Model 45 is the first paperless executive jet, designed entirely on screen by CAD/CAM. This is an especially formidable accomplishment given that it required the coordination of three different major component manufacturers in three different countries on two continents and communication between several

The wing arrives from De Havilland-Canada...

different software programs. In view of these challenges, the design process and prototype construction have gone remarkably smoothly.

The design makes extensive use of machined structures to reduce parts count, and a high degree of automation is utilized to streamline the production process. Because of the reliability of the design techniques, the Model 45 prototype was built on production tooling (just like the original Model 23, but for

entirely different reasons). In many instances in the production process, CAD/CAM files are directly loaded down into numerically controlled milling equipment that automatically makes the parts. This technique has allowed a 50 percent parts reduction compared to earlier construction methods. The forward fuselage bulkhead, for example, is milled from a single piece of metal in eight hours, compared to the conventional alternative that would require 75

The Learjet 45 fuselage arrives from Short's plant in Northern Ireland...

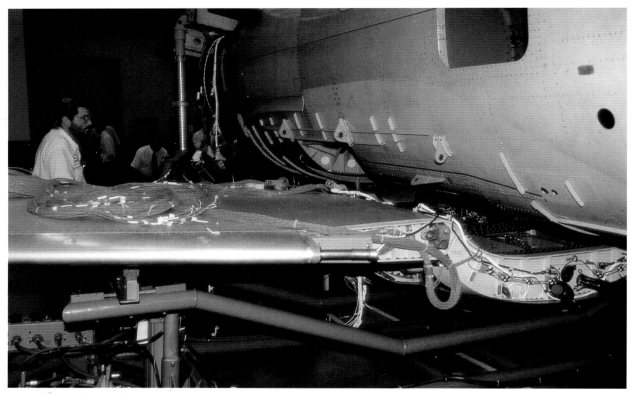

...and the two are a perfect fit.

parts and would take eight days to build. Even more impressive single-part milled components are the massive forward spar center sections. In the case of the wing, 80 percent of the parts are milled out of aluminum billets by numerically controlled machines. Learjet has experienced excellent results with the state-of-the-art design and construction process. The prototype parts rejection rate on the fuselage amounted to only 1.3 percent compared to typical parts rejection rates of up to 20 percent on prototypes built using earlier techniques.

The first big test for the Model 45 came in late 1994, when the prototype Shorts fuselage and the De Havilland wings were both delivered to Wichita and bolted together. The fit was perfect.

The Model 45 made its first flight on October 7, 1995, 32 years after (to the day) that the first Learjet took to the air. Flown by Pete Reynolds and Jim Dwyer, the 2 hour flight was flawless.

As it gains service experience, there is every reason to believe that the Model 45 will live-up to the high standards set by the Model 23 so long ago and maintained by every Learjet since. Learjet has ultimately come out a winner in the intensely competitive and highly cyclical aviation industry because of the excellence of its product and the dedication of the people who build it. And with its current line-up of aircraft, Learjet is sure to carry on the heritage of Bill Lear's sleek little jet that forever changed the world of executive aviation.

The Learjet 45's glass cockpit is equipped with Honeywell's Primus 1000 integrated avionics system.

The prototype Learjet 45 takes shape in Wichita.

High over Wichita, Kansas, the Learjet 45 on its maiden flight. *Paul Bowen*

RIGHT
Learjet 45 has one of the roomiest cabins in its class.

Appendix

	Gross Weight	Passengers	Engines (2)
Model 23	12,500 lb	5-7	General Electric CJ-610
Model 24	13,000 lb	5-7	General Electric CJ-610
Model 25	15,000 lb	4-6	General Electric CJ-610
Model 28/29	15,000 lb	5-7	General Electric CJ-610
Model 35 A	18,000 lb	7-9	Garret TFE731-2
Model 36 A	18,300 lb	4-6	Garret TFE731-2
Model 55	21,250 lb	6-10	Garret TFE731-3
Model 31 A	16,500 lb	7-9	Garret TFE731-2
Model 60	23,100 lb	6-10	Pratt & Whitney PW305A
Model 45	19,500 lb	8-10	Garret TFE731-20

	Thrust (x2)	Max Cruise	Max. Op. Alt.	Range
Model 23	2,850 lb	Mach 0.82	45,000 ft	1,500 nm
Model 24	2,950 lb	Mach 0.82	45-51,000 ft	1,500 nm
Model 25	2,950 lb	Mach 0.82	45-51,000 ft	1,750 nm
Model 28/29	2,950 lb	Mach 0.81	51,000 ft	1,200/1,800 nm
Model 35 A	3,500 lb	Mach 0.81	45,000 ft	2,196 nm
Model 36 A	3,500 lb	Mach 0.81	45,000 ft	2,500 nm
Model 55	3,700 lb	Mach 0.81	51,000 ft	2,200 nm
Model 31 A	3,500 lb	Mach 0.81	51,000 ft	1,561 nm
Model 60	4,600 lb	Mach 0.81	51,000 ft	2,750 nm
Model 45	3,500 lb	Mach 0.81	51,000 ft	2,200 nm

Index